Title: "Emerald Odyssey: The Global Journey of a Timeless Gem"

Warm regards,

FAISAL JAMIL

I Always Give's Free Copies Need Your Feedback And

Reviews Keeps In Touch!

http://www.amazon.com/author/faisal.jamil

Email: faisaljamilauthor@gmail.com

About the author

Certainly! Faisal Jamil is a multifaceted individual with a diverse set of skills and experiences. With a strong foundation in computer knowledge since childhood, he has developed a deep understanding of technology that informs his work as a content writer. Faisal also possesses digital skills, which further enhance his abilities in various digital platforms and technologies.

Beyond his professional endeavors, Faisal Jamil has also excelled in the martial arts, particularly Shotokan Karate, where he achieved the prestigious rank of first Dan black belt. This achievement speaks to his dedication, discipline, and commitment to personal growth and mastery.

In his professional life, Faisal Jamil has carved out a successful career in sales management within the Fast Moving Consumer Goods (FMCG) sector. His roles in various FMCG companies have honed his skills in strategic planning, team leadership, and business development. Faisal's ability to drive sales and achieve targets has been instrumental in his career progression, showcasing his talent for identifying opportunities and delivering results.

Faisal Jamil is also deeply interested in business investment strategies, planning, and execution. His understanding of these areas has been key to his success in the business world, allowing him to make informed decisions and implement effective strategies. His ability to navigate the complexities of investment planning and execution has set him apart as a strategic thinker and a valuable asset in any business endeavor.

Overall, Faisal Jamil is a dynamic individual who combines his passion for technology, martial arts, sales management, digital skills, and business investment strategies to achieve success in diverse fields. His journey is a testament to his versatility, resilience, and continuous pursuit of excellence.

Yours Sincerely

FAISAL JAMIL

I Always Give's Free Copies Need Your Feedback And

Reviews Keeps In Touch!

https://www.amazon.com/author/faisal.jamil

Email: faisaljamilauthor@gmail.com

EMERALD

ODYSSEY

THE GLOBAL JOURNEY
OF A TIMELESS GEM

Table of Content

Preface

Welcome to **Emerald Odyssey: The Global Journey of a Timeless Gem**, a journey through the rich and multifaceted world of emeralds. This book aims to unravel the captivating story of these stunning gemstones, exploring their origins, cultural significance, economic impact, and future prospects.

Emeralds, with their enchanting green hue, have been admired and coveted for thousands of years. Their allure is not merely a result of their beauty but also of the deep historical and cultural roots that stretch across civilizations. From the ancient mines of Egypt to the contemporary jewelry houses of today, emeralds have been symbols of power, wealth, and mystique.

This book is divided into five chapters, each dedicated to a unique aspect of the emerald story:

Chapter 1: The Enigmatic Origins of Emeralds - We delve into the geological wonders that create emeralds. The intricate processes of heat, pressure, and time that form these gems deep within the Earth's crust are as fascinating as the stones themselves. We explore the earliest known emerald mines and their historical significance.

Chapter 2: Emeralds in Culture and Mythology - Emeralds have played a significant role in various cultures and mythologies. This chapter takes you through the rich tapestry of stories and beliefs that have surrounded emeralds, from the ancient Romans and Egyptians to the lore of the emerald tablet in alchemy.

Chapter 3: The Global Trade of Emeralds - The journey of emeralds from mines to markets is a tale of global commerce and intricate trade networks. We examine the roles of major producing countries like Colombia, Brazil, and Zambia, and discuss the economic and ethical challenges the industry faces today.

Chapter 4: The Role of Emeralds in Modern Jewelry - The timeless appeal of emeralds continues to influence contemporary jewelry design. We explore how modern jewelers are incorporating these gems into their creations, blending traditional craftsmanship with innovative techniques, and setting trends in the world of high fashion.

Chapter 5: The Future of Emeralds - As we look to the future, the emerald industry is at a crossroads. Technological advancements, the rise of lab-grown emeralds, and the imperative of sustainability are shaping the future of emerald mining and trade. This chapter discusses the potential impacts of these developments and the industry's commitment to ethical practices.

In writing this book, my goal has been to provide a comprehensive and engaging exploration of emeralds, capturing the essence of what makes these gemstones so extraordinary. Whether you are a gem enthusiast, a jewelry aficionado, or simply intrigued by the rich history and vibrant allure of emeralds, I hope this book offers you a deeper appreciation for these timeless gems.

Thank you for joining me on this journey through the captivating world of emeralds. May **Emerald Odyssey: The Global Journey of a Timeless Gem** inspire you with the beauty, history, and enduring significance of these remarkable stones.

Sincerely,

FAISAL JAMIL

INTRODUCTION

Welcome to **Emerald Odyssey: The Global Journey of a Timeless Gem**, a comprehensive exploration of one of the world's most captivating gemstones. Emeralds, with their vibrant green hue and rich history, have enchanted humanity for millennia. This book aims to take you on a journey through the fascinating world of emeralds, uncovering their origins, cultural significance, economic impact, and future prospects.

From ancient civilizations to modern times, emeralds have held a unique place in human history. They have been symbols of power, beauty, and mystery, revered for their supposed mystical properties and valued for their rarity and allure. The journey of an emerald, from its formation deep within the Earth's crust to its final setting in a piece of jewelry, is a tale of natural wonder and human ingenuity.

In **Chapter 1: The Enigmatic Origins of Emeralds**, we delve into the geological processes that create these stunning gems. You'll learn about the unique conditions required for their formation and the historical mines that have produced some of the finest emeralds in the world.

Chapter 2: Emeralds in Culture and Mythology takes you through the cultural and mythological significance of emeralds across different civilizations. Discover how these gems were revered by ancient Romans, Egyptians, and Indians, and explore the legends and lore that have surrounded emeralds for centuries.

The global trade of emeralds is a complex and fascinating subject, covered in **Chapter 3: The Global Trade of Emeralds**. This chapter examines the roles of key producing countries, the economic impact of emerald mining, and the challenges faced by the industry, including ethical concerns and the rise of synthetic emeralds.

In **Chapter 4: The Role of Emeralds in Modern Jewelry**, we explore how contemporary jewelers incorporate emeralds into their designs, blending traditional craftsmanship with innovative techniques. This chapter highlights the enduring allure of emeralds in fashion and high jewelry, showcasing iconic pieces and current trends.

Finally, **Chapter 5: The Future of Emeralds** looks ahead to the future of the emerald industry. Technological advancements in mining and processing, the impact of climate change, and the growing market for lab-grown emeralds are all examined. This chapter also discusses the integration of blockchain technology for ethical sourcing and the ongoing commitment to sustainability.

Throughout this book, you will gain a comprehensive understanding of the multifaceted role that emeralds play globally. Whether you are a gem enthusiast, a jewelry aficionado, or simply captivated by the allure of these precious stones, **Emerald Odyssey: The Global Journey of a Timeless Gem** offers a thorough and insightful guide to the enduring legacy and global significance of emeralds. Join us on this enchanting journey and discover the timeless beauty and mystery of emeralds.

Chapter 1

The Enigmatic Origins of Emeralds

Emeralds, known for their mesmerizing green hue, have captivated humanity for millennia. This chapter delves into the geological processes that create these stunning gemstones, tracing their journey from deep within the Earth's crust to the surface, where they have been cherished and admired throughout history.

The Geological Birth of Emeralds

Emeralds are a variety of the mineral beryl, which also includes other gemstones such as aquamarine and morganite. The vibrant green color of emeralds is due to the presence of trace amounts of chromium and, in some cases, vanadium. The formation of emeralds requires a unique set of geological conditions, making them rarer than many other gemstones.

Deep within the Earth's crust, at depths of up to several kilometers, beryllium and aluminum-rich minerals interact with chromium or vanadium-bearing rocks. This process typically occurs in regions with hydrothermal activity, where hot, mineral-rich fluids circulate through cracks and fissures in the Earth's crust. Over millions of years, these fluids cool and crystallize, forming the beautiful green crystals known as emeralds.

The formation of emeralds involves immense heat and pressure, which contribute to their durability and the unique inclusions that are often present within the stones. These inclusions, sometimes referred to as "jardin" (French for "garden"), are internal fractures or foreign materials that give each emerald its distinct character. Far from being considered flaws, these inclusions are often prized for their ability to give emeralds a unique identity and authenticity.

Ancient Egyptian Emerald Mines

The earliest known emerald mines date back to ancient Egypt, around 1500 BCE. These mines, located in the Eastern Desert near the Red Sea, were a significant source of emeralds for the ancient world. The most famous of these mines is known as Cleopatra's Mines, named after the legendary queen who had a well-documented obsession with emeralds. Cleopatra's passion for the gem cemented its status as a symbol of power, beauty, and wealth.

Emeralds from Cleopatra's Mines were highly prized and often worn by royalty and high-ranking officials. The ancient Egyptians believed that emeralds had protective and healing properties, often using them in burial rites to ensure safe passage to the afterlife. The mines were extensively worked for several centuries before being abandoned, only to be rediscovered and explored again in the 19th century.

Colombian Emeralds: The World's Finest

While the Egyptian mines were historically significant, the most renowned emeralds come from Colombia. Discovered by Spanish conquistadors in the 16th century, Colombian emeralds are considered some of the finest in the world due to their exceptional color and clarity. The Muzo, Chivor, and Coscuez mines are the most famous and productive, each with a rich history of mining and trade.

The Spanish conquest of the Americas brought Colombian emeralds to the attention of Europe, where they quickly became highly sought after by royalty and the wealthy elite. The indigenous Muzo people, who had been mining emeralds long before the arrival of the Spanish, resisted the conquest fiercely but ultimately lost control of their valuable resources. The conquistadors exploited these mines extensively, sending vast quantities of emeralds back to Europe.

Colombian emeralds have continued to dominate the global market, with their deep green color and minimal inclusions setting the standard for quality. Modern mining techniques have allowed for more efficient extraction and processing, ensuring that Colombian emeralds remain a key player in the global gemstone industry.

The Tales of Conquistadors and Indigenous Miners

The rich history of emerald mining is filled with tales of adventure, conflict, and discovery. The Spanish conquistadors, driven by greed and the promise of unimaginable wealth, embarked on perilous journeys into the heart of the South American jungles. Their encounters with indigenous miners, who had a deep spiritual connection to the emeralds, were often marked by violence and exploitation.

Despite the tumultuous history, the legacy of these mines has endured, with emeralds from Colombia, Egypt, and other historical sites continuing to be prized by collectors and jewelers around the world. The chapter concludes by setting the stage for understanding the global journey of emeralds, from their ancient origins to their modern significance in the gemstone market.

In summary, the enigmatic origins of emeralds are a testament to the incredible geological processes that create these stunning gemstones. Their rich history, spanning from ancient Egypt to modern Colombia, is a story of human fascination, cultural significance, and enduring beauty. As we explore the journey of emeralds in subsequent chapters, we will uncover the many facets of their global role and impact.

Chapter 2

Emeralds in Culture and Mythology

Emeralds have been revered not only for their beauty but also for their supposed mystical properties. This chapter explores the cultural and mythological significance of emeralds across different civilizations, highlighting how these gemstones have transcended mere adornment to become powerful symbols in human history.

The Mystical Allure of Emeralds in Ancient Rome

In ancient Rome, emeralds were associated with Venus, the goddess of love and beauty. Roman writers like Pliny the

Elder praised emeralds for their vibrant color, which he believed was soothing to the eye. Pliny noted that emeralds were the only gemstones that delighted the eye without tiring it, a testament to their unique allure.

The Romans believed that emeralds had protective qualities, capable of warding off evil spirits and enhancing the well-being of their wearers. Emeralds were also thought to have a calming effect, promoting peace and harmony. These beliefs made emeralds popular among the Roman elite, who often wore them in jewelry and used them in various decorative items.

Emeralds in Ancient Egypt: Symbols of Eternal Life

The ancient Egyptians held emeralds in high regard, associating them with protection and healing. They believed that emeralds had the power to bestow eternal youth and protect against illness. Emeralds were often buried with the dead to ensure safe passage to the afterlife, reflecting their importance in funerary practices.

Cleopatra, the last Pharaoh of Egypt, was famously enamored with emeralds. She adorned herself with emerald jewelry and even claimed ownership of all emerald mines in Egypt. Her association with emeralds cemented their status as symbols of power and beauty. Egyptian emeralds were also used in religious ceremonies and as offerings to the gods, further underscoring their spiritual significance.

The Rich Tradition of Emeralds in India

In India, emeralds have long been considered symbols of wealth and prosperity. They are often featured in royal jewelry, signifying status and opulence. Indian maharajas and royalty adorned themselves with emeralds, believing them to bring good fortune and protect against negative influences.

Emeralds are also integral to traditional Indian medicine, or Ayurveda, where they are believed to have healing properties. They are thought to improve mental clarity, enhance communication, and promote emotional balance. The deep green color of emeralds is associated with the heart chakra, symbolizing love and compassion.

The Lore of the Emerald Tablet

One of the most fascinating aspects of emerald mythology is the legend of the Emerald Tablet, attributed to Hermes Trismegistus. The Emerald Tablet is said to contain the secrets of alchemy and the universe, written in cryptic verses. It is considered one of the most important texts in Western esotericism and has influenced a wide range of mystical and philosophical traditions.

The tablet's association with emeralds underscores the gemstone's mystical allure and its connection to ancient wisdom. The legend suggests that the Emerald Tablet holds the key to understanding the fundamental principles of the universe, making emeralds symbols of knowledge and enlightenment.

Emeralds in Mythology and Folklore

Beyond specific cultures, emeralds appear in a variety of mythologies and folklore around the world. In Greek mythology, emeralds were believed to be the tears of the goddess Isis, reflecting their association with love and sorrow. In medieval Europe, emeralds were thought to have the power to reveal truth and protect against enchantments, often used by clergy and nobility to ward off evil.

In South America, indigenous cultures revered emeralds long before the arrival of Europeans. The Muzo people of Colombia considered emeralds sacred, believing they were gifts from their gods. The legend of Fura and Tena, two lovers turned into mountains who shed emerald tears, is a poignant example of the deep cultural and spiritual significance of emeralds in this region.

Modern Cultural Significance

Today, emeralds continue to be revered for their beauty and symbolism. They are often associated with renewal and rebirth, thanks to their vibrant green color reminiscent of spring and new growth. Emeralds are also the birthstone for May, symbolizing love and success.

In contemporary jewelry design, emeralds are celebrated for their versatility and timeless appeal. They are frequently featured in high-end fashion and celebrity engagements, maintaining their status as symbols of luxury and elegance. The enduring fascination with emeralds across cultures and history underscores their unique place in the world of gemstones.

In conclusion, the cultural and mythological narratives surrounding emeralds highlight their significance beyond mere adornment. From ancient Rome and Egypt to India and beyond, emeralds have been powerful symbols of love, protection, wealth, and wisdom. Their mystical allure and timeless beauty continue to captivate and inspire, making them one of the most cherished gemstones in human history.

Chapter 3

The Global Trade of Emeralds

The trade of emeralds has a complex and fascinating history that spans continents and centuries. This chapter examines the journey of emeralds from mines to markets, detailing the roles of various countries in the global trade network. The emerald trade is a global enterprise, with Colombia, Brazil, and Zambia being significant players. This chapter will also address the economic impact of emerald mining on local communities and the global market, the challenges faced by the industry, including ethical concerns, and the efforts towards sustainable and fair-trade practices. Additionally, the rise of synthetic emeralds and their place in the market provides an interesting perspective on the evolving trade dynamics.

The Historical Journey of Emerald Trade

Emeralds have been traded since ancient times, with the earliest known transactions dating back to ancient Egypt. Cleopatra's famed mines supplied the ancient world with these precious stones, and trade routes extended across the Mediterranean and into Asia. The Roman Empire, with its vast network of trade routes, facilitated the spread of emeralds throughout Europe.

During the Age of Exploration in the 16th century, Spanish conquistadors discovered vast emerald deposits in Colombia. This discovery significantly shifted the center of emerald production and trade to the New World. The Spanish conquest of the Muzo, Chivor, and Coscuez mines in Colombia provided Europe with a steady supply of high-quality emeralds, which were eagerly sought after by European royalty and the wealthy elite.

Colombia: The Epicenter of Emerald Quality

Colombia remains the dominant player in the global emerald trade, known for producing some of the highest quality emeralds. The country's emeralds are renowned for their deep green color and exceptional clarity. The Muzo, Chivor, and Coscuez mines continue to be prolific sources of these coveted gemstones.

Colombian emeralds are often considered the gold standard in the industry, commanding higher prices than emeralds from other regions. The country's emerald industry plays a crucial role in its economy, providing employment and generating significant revenue. However, the industry also faces challenges, including illegal mining, smuggling, and the need for sustainable mining practices.

Emerging Players: Brazil and Zambia

While Colombia has long been synonymous with emeralds, other countries have emerged as significant sources in recent decades. Brazil and Zambia, in particular, have become major players in the global emerald trade.

Brazil: Brazilian emeralds were discovered in the early 20th century, and the country has since become a significant supplier. Brazilian emeralds are known for their bright green color, often with a slight blue hue. The country's major mining regions, such as Minas Gerais and Bahia, produce emeralds that are highly valued in the market. Brazil's emerald industry has benefited from modern mining techniques and a strong focus on sustainable practices.

Zambia: Zambia's emerald industry began to gain prominence in the late 20th century. The Kagem mine, the largest emerald mine in the world, is located in Zambia's Copperbelt Province. Zambian emeralds are known for their rich, saturated green color and high transparency. The country's emerald industry has grown rapidly, contributing significantly to the local economy. Efforts to implement ethical mining practices and community development projects have helped improve the industry's reputation on the global stage.

Economic Impact and Community Benefits

The emerald trade has a profound economic impact on producing countries. In Colombia, Brazil, and Zambia, the industry provides employment for thousands of people, from miners to those involved in cutting, polishing, and trading the gemstones. The revenue generated from emerald exports contributes to national economies and supports local communities.

However, the economic benefits of emerald mining are often accompanied by challenges. Illegal mining operations can lead to environmental degradation, poor working conditions, and loss of revenue for legitimate miners and governments. Addressing these issues requires robust regulatory frameworks and collaboration between governments, industry stakeholders, and local communities.

Ethical Concerns and Sustainable Practices

The emerald industry, like other gemstone industries, faces ethical concerns related to environmental impact, labor practices, and the fair distribution of profits. In response, there has been a growing movement towards sustainable and fair-trade practices in emerald mining.

Efforts to promote ethical emerald mining include:

1: Certification Schemes:

Initiatives such as the Responsible Jewellery Council (RJC) and the Fairtrade Gold and Precious Metals certification aim to ensure that emeralds are sourced responsibly. These schemes promote transparency and traceability in the supply chain, helping consumers make informed choices.

2: Community Development Projects:

Mining companies are increasingly investing in community development projects, such as education, healthcare, and infrastructure improvements. These initiatives aim to improve the quality of life for local communities and create a more positive impact from emerald mining activities.

3: Environmental Stewardship:

Sustainable mining practices focus on minimizing environmental impact through measures such as land reclamation, reducing water usage, and managing waste. Companies that prioritize environmental stewardship are better positioned to meet the growing demand for ethically sourced emeralds.

The Rise of Synthetic Emeralds

The emergence of synthetic emeralds has added a new dimension to the global trade dynamics. Synthetic emeralds, created in laboratories using advanced technologies, offer a cost-effective alternative to natural emeralds. They possess similar physical and chemical properties to natural emeralds, making them nearly indistinguishable to the untrained eye.

Synthetic emeralds have gained popularity due to their affordability and ethical appeal. They are free from the environmental and social issues associated with mining, making them an attractive option for eco-conscious consumers. However, the rise of synthetic emeralds also presents challenges for the natural emerald market, as it requires educating consumers about the differences and value of natural versus synthetic gemstones.

In conclusion, the global trade of emeralds is a complex and dynamic industry with a rich history and significant economic impact. The roles of countries like Colombia, Brazil, and Zambia highlight the geographical diversity and evolving nature of the emerald market. Addressing ethical concerns and promoting sustainable practices are essential for the industry's future. The rise of synthetic emeralds adds an interesting perspective to the trade, offering both opportunities and challenges. Understanding these facets of the emerald trade provides a comprehensive view of its global significance and enduring allure.

Chapter 4

The Role of Emeralds in Modern Jewelry

Emeralds have a timeless appeal that continues to influence contemporary jewelry design. This chapter explores how modern jewelers incorporate emeralds into their creations, blending traditional craftsmanship with innovative techniques. Renowned jewelry houses like Cartier, Bulgari, and Tiffany & Co. have created iconic pieces featuring emeralds, often pushing the boundaries of design and artistry. The chapter also highlights the trends in emerald jewelry, from statement pieces worn by celebrities on red carpets to custom-designed engagement rings that symbolize eternal love. The enduring allure of emeralds in fashion and high jewelry underscores their status as a gemstone that transcends time and trends.

The Timeless Appeal of Emeralds in Jewelry Design

Emeralds, with their captivating green hue, have been cherished for centuries and continue to be a favorite among modern jewelers. The vibrant color of emeralds, often associated with renewal and growth, makes them a versatile choice for a variety of jewelry pieces. Their unique inclusions, known as "jardin," add character and individuality to each stone, making every emerald piece truly one-of-a-kind.

Contemporary jewelers embrace the challenge of working with emeralds, incorporating them into designs that highlight their natural beauty. Traditional craftsmanship is often combined with modern techniques to create pieces that are both classic and innovative. Jewelers skillfully cut and set emeralds to maximize their brilliance and showcase their rich color, ensuring that each piece stands out.

Iconic Jewelry Houses and Their Emerald Creations

Several renowned jewelry houses have a long history of creating iconic pieces featuring emeralds. These houses have set the standard for high jewelry, blending exquisite design with exceptional craftsmanship.

Cartier: Known for its luxurious and innovative designs, Cartier has created some of the most famous emerald jewelry pieces in history. One of the most notable is the "Panthère" collection, which often features emeralds as the eyes of the iconic panther motif. Cartier's use of emeralds in combination with other precious stones and metals results in pieces that are both timeless and contemporary.

Bulgari: Bulgari is celebrated for its bold and colorful designs, often incorporating emeralds in striking combinations with other gemstones. The Italian jewelry house's use of cabochon-cut emeralds, paired with diamonds and sapphires, creates vibrant and dynamic pieces that stand out in the world of high jewelry. Bulgari's ability to blend traditional elements with modern aesthetics has made it a favorite among collectors and celebrities alike.

Tiffany & Co.: Tiffany & Co. has a long-standing reputation for creating elegant and refined jewelry. The American jewelry house's use of emeralds in its designs reflects its commitment to quality and sophistication. Tiffany's engagement rings and high jewelry pieces often feature emeralds, set in designs that highlight the stone's natural beauty and elegance.

Trends in Emerald Jewelry

Emerald jewelry continues to be a popular choice for a variety of occasions, from red carpet events to everyday wear. Several trends have emerged in recent years, showcasing the versatility and enduring appeal of emeralds.

Statement Pieces: Emeralds are often featured in bold, statement pieces worn by celebrities on red carpets and at high-profile events. These pieces, which can include necklaces, earrings, and bracelets, highlight the gemstone's vibrant color and luxurious appeal. Celebrities such as Angelina Jolie and Elizabeth Taylor have famously worn emerald jewelry, further cementing the stone's status as a symbol of glamour and elegance.

Custom-Designed Engagement Rings: Emeralds have become an increasingly popular choice for engagement rings, offering a unique and distinctive alternative to traditional diamonds. Custom-designed emerald engagement rings symbolize eternal love and are often chosen for their rich color and unique inclusions. These rings can be tailored to reflect personal style and preferences, making them a meaningful and cherished choice for couples.

Vintage and Antique Styles: The popularity of vintage and antique jewelry has also contributed to the enduring appeal of emeralds. Many modern jewelers draw inspiration from historical designs, creating pieces that evoke the elegance and charm of bygone eras. Art Deco and Victorian-inspired designs, featuring emeralds set in intricate and detailed settings, are particularly popular among collectors and jewelry enthusiasts.

The Enduring Allure of Emeralds in Fashion and High Jewelry

Emeralds continue to be a staple in the world of fashion and high jewelry, admired for their timeless beauty and versatility. Their rich green color complements a wide range of styles and can be paired with various metals and gemstones to create unique and stunning pieces.

In high jewelry, emeralds are often used in combination with diamonds, sapphires, and other precious stones to create pieces that are both luxurious and sophisticated. The intricate craftsmanship and attention to detail that go into creating high jewelry pieces ensure that emeralds are showcased to their full potential.

Emeralds also play a significant role in fashion jewelry, where they are used to create more accessible and contemporary designs. From minimalist pieces to bold, statement jewelry, emeralds add a touch of elegance and sophistication to any ensemble.

In conclusion, emeralds hold a special place in modern jewelry design, continuing to captivate and inspire both jewelers and wearers. The blending of traditional craftsmanship with innovative techniques allows for the creation of timeless and unique pieces that highlight the natural beauty of emeralds. Renowned jewelry houses like Cartier, Bulgari, and Tiffany & Co. have set the standard for high jewelry, while trends in statement pieces and custom-designed engagement rings reflect the gemstone's enduring appeal. The allure of emeralds in fashion and high jewelry underscores their status as a gemstone that transcends time and trends, remaining a cherished and admired choice for generations.

Chapter 5

The Future of Emeralds

As we look to the future, the emerald industry faces both challenges and opportunities. This chapter discusses the technological advancements in gemstone mining and processing, which promise to make emerald extraction more efficient and environmentally friendly. Innovations such as blockchain technology are being explored to ensure the traceability and ethical sourcing of emeralds. The chapter also examines the potential impacts of climate change on emerald mining regions and the steps being taken to mitigate these effects. Additionally, it considers the growing market for lab-grown emeralds and how they might shape the future of the industry. The fusion of tradition and technology, along with a commitment to sustainability, will play a crucial role in the continued global significance of emeralds.

Technological Advancements in Gemstone Mining and Processing

The emerald mining industry is witnessing significant technological advancements aimed at improving efficiency and reducing environmental impact. These innovations are transforming traditional mining methods and paving the way for a more sustainable future.

Automated and Remote Mining:

The use of automated machinery and remote-controlled equipment is becoming increasingly common in emerald mines. These technologies enhance precision and safety, allowing for more accurate extraction while minimizing damage to the surrounding environment. Automated systems can also reduce labor costs and improve the overall efficiency of mining operations.

Advanced Sorting and Grading Techniques:

Innovations in sorting and grading technology are improving the quality control process for emeralds. Advanced optical and laser sorting systems can quickly and accurately categorize gemstones based on their color, clarity, and size. These technologies help ensure that only the highest quality emeralds reach the market, enhancing their value and desirability.

Eco-Friendly Processing Methods:

The development of environmentally friendly processing techniques is a major focus for the emerald industry. Water recycling systems, non-toxic chemical treatments, and reduced energy consumption are among the methods being implemented to minimize the environmental footprint of emerald processing. These practices not only protect natural resources but also align with growing consumer demand for sustainably sourced products.

Blockchain Technology and Ethical Sourcing

Blockchain technology is emerging as a powerful tool for ensuring the traceability and ethical sourcing of emeralds. By providing a secure and transparent digital ledger, blockchain can track the journey of an emerald from the mine to the market, offering verifiable proof of its origin and ethical production practices.

Transparency and Accountability:

Blockchain technology allows for the creation of an immutable record of each step in the emerald supply chain. This transparency ensures that consumers can trust the provenance of their gemstones, reducing the risk of purchasing conflict minerals or stones sourced through unethical practices.

Consumer Confidence:

As awareness of ethical sourcing grows, consumers are increasingly seeking assurance that their purchases align with their values. Blockchain-enabled traceability can provide this assurance, boosting consumer confidence and potentially increasing demand for ethically sourced emeralds.

Industry Collaboration:

The adoption of blockchain technology requires collaboration among miners, processors, distributors, and retailers. Industry-wide standards and protocols are being developed to facilitate the implementation of blockchain systems, ensuring that all stakeholders benefit from enhanced transparency and traceability.

Climate Change and Its Impact on Emerald Mining

Climate change poses a significant threat to emerald mining regions, with potential impacts on both the environment and the communities that depend on mining for their livelihoods. Addressing these challenges requires proactive measures to mitigate the effects of climate change and build resilience.

Environmental Changes:

Shifts in temperature and precipitation patterns can affect the geological conditions necessary for emerald formation. Changes in water availability, soil composition, and vegetation cover can disrupt mining operations and impact the quality of gemstones.

Community Resilience:

Many emerald mining communities are vulnerable to the effects of climate change, including extreme weather events and resource scarcity. Initiatives to enhance community resilience, such as infrastructure improvements, diversified income sources, and access to education and healthcare, are critical for sustaining livelihoods and ensuring the long-term viability of mining regions.

Sustainable Practices:

Implementing sustainable mining practices is essential for mitigating the environmental impact of emerald extraction. Reforestation projects, soil conservation efforts, and the adoption of renewable energy sources are among the strategies being employed to reduce the carbon footprint of mining activities and protect ecosystems.

The Rise of Lab-Grown Emeralds

The growing market for lab-grown emeralds presents both opportunities and challenges for the traditional gemstone industry. Lab-grown emeralds, created using advanced technological processes, offer a sustainable and affordable alternative to natural stones.

Sustainable Production:

Lab-grown emeralds are produced in controlled environments, eliminating the need for mining and reducing the associated environmental and social impacts. The production process requires fewer resources and generates less waste, making lab-grown emeralds an attractive option for eco-conscious consumers.

Affordability and Accessibility:

The lower production costs of lab-grown emeralds make them more affordable than their natural counterparts. This accessibility opens up new market segments and allows a wider range of consumers to enjoy the beauty of emeralds.

Market Dynamics:

The rise of lab-grown emeralds is influencing market dynamics, with implications for pricing, consumer preferences, and industry practices. While some consumers may prefer the authenticity and rarity of natural emeralds, others are drawn to the ethical and economic advantages of lab-grown stones. The coexistence of both types of emeralds in the market highlights the importance of consumer education and informed decision-making.

The Fusion of Tradition and Technology

The future of emeralds lies in the successful integration of traditional craftsmanship with modern technology. This fusion enhances the appeal of emeralds while ensuring that their production aligns with contemporary values of sustainability and ethical sourcing.

Craftsmanship and Innovation:

Jewelers and designers are increasingly blending traditional techniques with innovative approaches to create unique and captivating pieces. This synergy allows for the preservation of heritage craftsmanship while embracing new possibilities in design and production.

Consumer Trends:

As consumers become more discerning and values-driven, the demand for responsibly sourced and beautifully crafted emeralds is expected to grow. Jewelers who prioritize transparency, sustainability, and quality will be well-positioned to meet this demand and maintain the enduring allure of emeralds.

Sustainable Future:

The commitment to sustainability in the emerald industry encompasses environmental stewardship, ethical practices, and community development. By adopting a holistic approach, the industry can ensure the continued global significance of emeralds while contributing to a more equitable and sustainable world.

Conclusion

By the end of this book, readers will have a comprehensive understanding of the multifaceted role that emeralds play globally, from their geological origins to their cultural significance, economic impact, and future prospects. The enduring allure of emeralds, coupled with advancements in technology and a commitment to sustainability, ensures that these captivating gemstones will continue to be cherished and admired for generations to come.

The End!

www.ingramcontent.com/pod-product-compliance
Lightning Source LLC
Chambersburg PA
CBHW040758240526
45474CB00008B/105